FUT... ...S INTELLIGENCE

*Assessing Intelligence Support
to Three Army Long-Range
Planning Communities*

JOHN E. PETERS
ERIC V. LARSON
JAMES A. DEWAR

Prepared for the
United States Army

Arroyo Center

RAND

For more information on the RAND Arroyo Center, contact the Director of Operations, (310) 393-0411, extension 6500, or visit the Arroyo Center's Web site at http://www.rand.org/organization/ard/

This report highlights the principal lines of investigation and findings of a project entitled "Intelligence Support to Long-Range Planning," undertaken for the U.S. Army's Office of the Deputy Chief of Staff for Intelligence. The project examined intelligence support to the three main groups of Army long-range planners. The research was conducted in the Strategy and Doctrine Program of RAND's Arroyo Center, a federally funded research and development center sponsored by the United States Army. It should interest those with responsibilities either in futures intelligence, acquisition, force development, or other forms of long-range planning. The report should also prove useful for those involved in the Army's future force initiatives, Force XXI and the Army After Next.

CONTENTS

FIGURES

TABLES

SUMMARY

This project began as a review of intelligence support to Army long-range planning. The study team first sought to identify who the Army long-range planners were and, more specifically, to answer three questions: (1) How is intelligence support to long-range planning performed? (2) What does the current planning system require from intelligence? and (3) What are the prospects for commercial information management technologies to improve intelligence support to long-range planning? Answering these questions involved reviewing intelligence documents and production records, doing case studies of three large and highly capitalized firms, conducting workshops to which planners and intelligence staff were invited, and canvassing consumers of futures intelligence—intelligence that attempts to anticipate future circumstances, conditions, and even foreign military postures.

Three groups of planners ultimately emerged as primary consumers of futures intelligence: the strategic planners, the acquisition community, and the force developers. The needs of these planning groups were in some ways quite distinct. The acquisition and force development people had significant needs for detailed, point estimates, while the strategic planners and their army reinvention colleagues were less interested in such specifics. Despite their differences, all these planning constituencies shared a common approach to planning, sometimes called capabilities-based planning. Capabilities-based planning tends to emphasize technology and, in some applications, to emphasize what U.S. technology can make available for Army application.

As the study progressed, it became clear that some intelligence consumers—principally threat integration staff officers who provide interface between intelligence and specific acquisition programs—were unhappy with the responsiveness of Army intelligence. Some worried about the quality of the intelligence, and others doubted that the U.S. Army's Office of the Deputy Chief of Staff for Intelligence (ODCSINT) was customer oriented enough. Specific concerns and issues varied among the planners, but a common point of friction arose from the fact that intelligence is threat-based while the planners' approaches were capabilities-based. More specifically, because the planners were more narrowly focused, they did not appreciate the potential for trouble that lurked in some of the assumptions implicit in their work, or Army intelligence's value for helping to identify and develop responses to these dangers.

Besides diagnosing current points of friction between planners and their intelligence support, the study group examined possible ways that ODCSINT might improve its support to the planners. Three separate avenues emerged, one technical, one methodological, one conceptual.

The technical route is communications and connectivity technology that has the potential to improve connectivity between ODCSINT and its customers. The Army is already deeply invested in this field and the first benefits to Army intelligence have begun to appear, such as Intelink, a secure Internet-like computer system. But ODCSINT must act to ensure that the Army's information technology (IT) technical architecture continues to develop in ways that will contribute greater connectivity and linkages between Army intelligence and its customers. This means ODCSINT must work closely with the Director for Information Systems of Command, Control, Communications, and Computers (DISC4) and others to understand customers' needs and plan the necessary systems.

The methodological avenue leads to improved planner support by monitoring the important assumptions implicit and explicit in the planners' approaches to their respective tasks. Since the planners tend to focus on the narrow aspects of capabilities-based planning, the broader, more encompassing threat orientation of Army intelligence can help protect the planners by identifying vulnerable assumptions before they fail and helping the planners craft appropriate responses.

The conceptual avenue leads to improved planner support by recognizing that ODCSINT's products are not solely its reports, but also its expertise, resident in human capital—its regional and functional experts. ODCSINT can improve its support to planners by providing for sustained interaction of its experts with its customers. Communications technology plays a role here, but the more important task is to make sure that ODCSINT continues to develop high-quality experts with sound reputations among Army planners and in the intelligence field.

ACKNOWLEDGMENTS

This project benefited from the generous support of many people, including those who participated in the workshops, responded to interviews and questionnaires, and generally assisted the research team in gathering information. Several deserve special mention, including Sheryl Root of Hewlett-Packard, Thomas Grumm of General Motors, Jon Parker of Boeing, and Alan Goldman, Scott Mingledorff, Bob O'Connell, and Art Peterson, all from the National Ground Intelligence Center. LTC Lonnie Henley, Patrick Neary, and Eric Vardac of the DCSINT staff, and COL Tom Molino and LTC(P) Tim Daniels from the DCSOPS staff were also instrumental in the success of the project. Colleagues Myron Hura and Bruce Pirnie provided thoughtful and insightful reviews of an earlier version of this report.

AAN	Army After Next
ACQ	Acquisition
APINS	Army Priority Intelligence Needs Survey
AR	Army Regulation
AVICE	Assistant Vice Chief of Staff of the Army
C4I	Command, control, communications, computers, and intelligence
CIA	Central Intelligence Agency
CINC	Commander-in-Chief
DCSINT	Deputy Chief of Staff for Intelligence
DCSOPS	Deputy Chief of Staff for Operations and Plans
DIA	Defense Intelligence Agency
DISC4	Director for Information Systems of Command, Control, Communications, and Computers
DOD	Department of Defense
DODFIP	DOD Futures Intelligence Program
DODIPP	DOD Intelligence Production Plan
FD	Force development
FIO	Foreign intelligence office

GNP	Gross national product
IT	Information Technology
ODCSINT	Office of the Deputy Chief of Staff for Intelligence
ODCSOPS	Office of the Deputy Chief of Staff for Operations and Plans
OSD	Office of the Secretary of Defense
PEG	Program evaluation group
STAR	System threat assessment reports
TRADOC	Training and Doctrine Command
WMD	Weapons of mass destruction

Chapter One

INTRODUCTION

This review of intelligence support to long-range planning began by examining each of the three planning activities that constitute Army long-range planning and their intelligence needs: (1) those broader activities of the Strategy, Plans, and Policy Directorate within DCSOPS and the major commands (MACOMS) that ensure the Army can fulfill its role in executing the National Military Strategy (hereinafter referred to simply as strategic planning), (2) acquisition (ACQ), which includes the formal members of the acquisition community and the constellation of laboratories, arsenals, and contractors that supports it, and (3) force development (FD), the extended collection of force planners and force integration experts who help to chart the transition of today's Army into tomorrow's force. This report summarizes the results of the review and offers recommendations based on our analysis.

RESEARCH QUESTIONS AND APPROACH

Three questions guided the inquiry: (1) How is intelligence support to long-range planning performed? (2) What does the current planning system require from intelligence? and (3) What are the prospects for commercial information management technologies to improve intelligence support to long-range planning? These questions are important because their answers suggest collectively the demands that long-range planning makes on Army intelligence. The study also considers the degree to which Army intelligence is prepared to satisfy planner needs and what adjustments to intelligence support might be appropriate. It is also worth noting what this study does not do. It does not attempt a comprehensive survey of intelligence consumers to gauge their satisfaction with specific intelligence products. The fiscal year 1994 and 1995 Army Priority Intelligence Needs Survey (APINS) studies have already done this job.

Several means of investigation combined to answer the study's questions. Surveys and interviews shed light on the value of some recent intelligence products and planning support. Case studies, especially of acquisition initiatives, illuminated the role of Army intelligence. A pair of workshops offered the opportunity to ask a limited number of planners explicitly what they needed to know and afforded planners and intelligence staff the time to talk directly. None of the investigatory methods used—surveys, interviews, or workshops—were intended to be statistically significant, nor aimed at establishing averages of data. Rather, they were intended to help the research team establish a range of opinions and impressions about intelligence support to long-range planning from an eclectic sample of individuals involved either as producers or consumers of that intelligence. The project also reviewed Army and Department of Defense (DOD) regulations and instructions, a variety of intelligence products, and the DOD Futures Intelligence Program (DODFIP). DODFIP was important because, according to the Defense Intelligence Agency (DIA), while it accounts for less than 1 percent of total DOD intelligence production, it has major impact upon (though not exclusive control of) futures work and estimative intelligence.[1]

Next, research turned toward information management, or more specifically to communications and connectivity, to see what the commercial world might offer that could improve intelligence support to planning. For this task, we began with a review of the published literature on strategic planning and the relationship between strategic planning and investments in information technology (IT). We then identified a small number of industries that shared some salient features with the Army. We focused on industries and firms with large-capitalization and robust research and development (R&D), but we also decided to be sure to include industries that varied in one important respect—product focus and planning horizon,

---

[1]Interview with Christine McKeown, TA-1 Division Chief, DOD Futures Intelligence Program, Defense Intelligence Agency, June 18, 1997.

as measured by typical product development and life cycle—to see if variations here made any difference in the way they conducted their planning. By synthesizing the results from the various investigative processes, the project proposes multiple answers to the study's main questions.

Finally, in one of the early workshops for the project, several participants asserted that Army intelligence was "broken." They argued that the system was not organized, equipped, or supplied with incentives to support long-range planning. Our research plan addressed these charges by looking for evidence throughout the investigation that would help confirm or disconfirm these claims.

REPORT ORGANIZATION

The remainder of this report is organized into five chapters.

- Chapter Two provides background necessary to grasp fully the prospects for improved intelligence support to long-range planning. It briefly examines the historical performance of both Army long-range planning and intelligence. The chapter also considers another important influence on intelligence support: the effects of the DOD Futures Intelligence Program on the way Army intelligence operates.

- Chapter Three looks carefully at the characteristics and needs of the three groups of long-range planners: strategic planners, ACQ, and FD.

- Chapter Four takes a look at communications and connectivity technology and the prospects that it can improve intelligence support to long-range planning.

- Chapter Five explores the prospects for Army intelligence to satisfy long-range planners' intelligence needs. The chapter considers systemic and methodological issues that might cast doubt on Army intelligence's ability to support the planners fully.

- Chapter Six concludes the report by summing up the salient observations from each of the prior chapters and offering recommendations on how Army intelligence might proceed with efforts to strengthen its capacity to support long-range planning.

Chapter Two

BACKGROUND

HISTORICAL PERSPECTIVE

Intelligence support to long-range planning is a relatively new endeavor for Army intelligence.[1] Until the Cold War, most efforts went toward current intelligence—studies of the enemy, weather, and terrain Army forces were likely to encounter. Prior to World War II, strong parochial and isolationist influences caused the Army to weigh factors other than intelligence—or strategic planning—in shaping Army decisions about technology, doctrine, and force structure. For example, the Army endorsed Plan Orange, the plan for war in the Pacific, not because it was congruent with its own intelligence estimates (it had none on the subject), but because Plan Orange would provide the budgetary support the Army sought for key programs.[2] Force development proceeded largely as if the Army had no interest in foreign capabilities and developments. Indeed, as late as 1940 the Army's force design was predicated upon committing the Army to a single theater, despite the obvious Japanese and German military postures and dispositions suggesting that if war came, the United States would fight in two major theaters. And rather than taking the long view, technical intelligence focused on the current

---

[1]Long-range planning usually means planning conducted independently of or beyond the reach of the Army's current budget and program constraints, where new priorities and budget decisions can be considered. Strategic planning generally means planning that is very important to the Army's ability to fulfill its responsibilities under the National Security Strategy and may or may not consider some distant, future time frame.

[2]Williamson Murray and Allan R. Millett (eds.), *Military Innovation in the Interwar Period* (Cambridge University Press, 1996), p. 58.

term, trying to understand the performance capabilities of foreign systems in the inventory at the time. Little effort was made to extrapolate or forecast future capabilities.[3] Although U.S. acquisition efforts responded, especially in wartime, to current intelligence reporting, long-term research and development efforts took relatively little notice of foreign progress. As Rosen observed,

> The overall picture of American research and development in the period from 1930 to 1955 is one of technical innovation largely unaffected by the activities of potential enemies, a rather self-contained process in which actions and actors within the military establishment were the main determinants of innovation... Military innovation is much less bound up with foreign military behavior or civilian invention than is ordinarily thought.[4]

In recent history, then, other factors besides intelligence or long-range planning predominated in shaping the Army's strategic, acquisition, and force development plans. Using intelligence to support these activities is a recent and historically novel idea.

Since World War II, intelligence support to long-range planning has become big business. As Table 1 illustrates, one of the Army's principal intelligence production facilities, the National Ground Intelligence Center, devotes about 85 percent of its efforts to "futures" intelligence—intelligence that seeks to estimate future capabilities 7, 10 and 20 years out in the future. Through much of the Cold War, the emphasis in futures intelligence was on "getting the Soviets right"—being able to forecast accurately how the Soviet ground forces would develop over time.

THE INFLUENCE OF LONG-RANGE PLANNING

The Cold War–vintage Army planning system as it evolved had a checkered history. It generally consisted of multiple tiers in which the planners derived much of their view from the Office of the Secre-

---

[3]See the United States Army in World War II series, *The Ordnance Department: Planning Munitions for War* (Office of the Center for Military History, Department of the Army, Washington, D.C., 1955), pp. 208–215.

[4]Stephen Peter Rosen, *Winning the Next War: Innovation and Modern Militaries* (Cornell University Press, 1991), p. 250.

Table 1

Distribution of National Ground Intelligence Center Production Assets

Category of Production	Total All Directorates (%)	Forces Directorate (%)	Systems Directorate (%)	Technologies Directorate (%)
Current	15	30	10	5
Futures	85	70	90	95
Global security forecast	1	4		
Regional security forecasts	6	18		
Scenarios, threat operational concepts	7	20		
Country forecasts	10	28		
Systems studies	44		75	60
Technology studies	17		15	35

SOURCE: Data provided by the U.S. Army National Ground Intelligence Center.

tary of Defense and Joint Staff's publications such as the Joint Long-Range Strategic Appraisal (JLRSA). Much of the Army planners' job was to distill the Joint Staff guidance for Army consumption and to flesh out the details of that guidance in Army terms. This allowed little latitude for viewing problems from an Army-unique perspective, since the planners were not authorized to stray from the basic findings of the source documents. An Army-unique perspective might identify issues or dangers that would not be obvious from the level of detail in a Joint Staff or OSD-prepared report. Moreover, there were tensions between the tiers in the planning system.

The Army Staff planners sought to offer the most specific guidance possible, while the Army Secretariat preferred to preserve as much of its own flexibility as possible. As a result, the Secretariat often sought to dilute the planners' guidance and make it as general as possible so it would not constrain the Secretariat's options.[5] In addition, the

---

[5]John E. Peters, *The U.S. Military: Ready for the New World Order?* (Greenwood Press: Westport, CT and London, 1993).

strategic planners were never fully integrated with the functional area planners or with the budget and program decisionmakers. The Army Staff's functional areas (the administrative sections within the staff across which the Army fulfills its Title 10 responsibilities to raise, train, equip, maintain, and sustain forces were distributed) had their own integral planners and plans for the future, with direct connection to budget and programmatic decisions. The strategic planners had limited portfolios that did not entitle them to significant influence over budget, program, and functional area issues. Indeed, throughout the Cold War, the Army focused on strategic *programming* (making decisions about the number and type of units in the force, their organization, and their equipment) rather than on strategic *planning* (the overall approach to the United States' security and the Army's role in it).[6] Therefore, intelligence support to their efforts had little influence because the planners themselves enjoyed little.

FORCE DEVELOPMENT

Unlike the other categories of long-range planning, force development has a lengthy history. Post–World War II major initiatives range from 1954's Atomic Field Army 1 through AirLand Battle Future in the late 1980s to today's Force XXI and AAN. Such reinvention actions began as responses to fairly near-term concerns (e.g., the emergence of the Soviet nuclear capability) but gradually aimed further into the future.[7]

FD is generally conceived of as involving interactions among doctrine, new equipment, and organization and structures. Army history suggests that at least five broad, general forces cause change: (1) the Army force design process, which relies routinely upon setting size "caps" as a part of its management process, (2) doctrinally

---

[6]The case for this point of view is argued persuasively in Carl H. Builder and James A. Dewar, "A Time for Planning? If Not Now, When?" *Parameters* (Summer 1994), pp. 4–15.

[7]For a review of the earliest post–World War II intiatives, see Major Glen R. Hawkins, *United States Army Force Structure and Force Design Initiatives 1939–1989*, U.S. Army Center for Military History (Advance Copy), 1991, pp. 12–13.

driven change, (3) new equipment and weapons, (4) new or changed threats, and (5) combat experience.[8]

Through most of the Cold War, FD's projects paid some attention to the threat, specifically, to forecast improvements in Soviet military capabilities. By the late 1980s, however, the threat role began to decline, and its representation in key documents decreased. For example, TRADOC, writing about Army 21 (a conceptual model of a future army) in 1986, devoted four pages to the threat. A U.S. Army TRADOC briefing on AirLand Battle Future in 1990 mentioned the threat only in passing; indeed, none of the subsequent points made in the presentation derived from problems posed by the threat. By 1994, the threat had all but vanished from a Louisiana Maneuvers progress report.[9]

ARMY INTELLIGENCE TODAY

Army intelligence's products tend to be derivative of intelligence produced by the other agencies that operate specific intelligence-collection disciplines. Army intelligence is often called "all-source," since it draws on finished intelligence and raw information including unclassified, open sources such as newspapers and television in producing its own reports. Because its products are based on a wide variety of sources that allow cross-checking and confirmation, Army intelligence may enjoy some protection from deliberate deception-and-denial programs run by foreign governments.

---

[8]See U.S. General Accounting Office, *Troop Reductions: Lessons Learned from Army's Approach to Inactivating the 9th Division*, GAO/NSIAD-92-78, June 1992, Major Robert A. Doughty, *The Evolution of U.S. Army Tactical Doctrine, 1946–1976* (Combat Studies Institute, Fort Leavenworth, KS, 1979), and Total Army Analysis FY 86–90 (TAA-90), Annex D, "Doctrinal requirements for nondivisional aviation, field artillery, and air defense artillery," U.S. Army Concepts Analysis Agency, November 1983.

[9]The Louisiana Maneuvers were an initiative of the early 1990s with the objective of modernizing the Army in the aftermath of the Cold War. See Headquarters, U.S. Army Training and Doctrine Command, "Army 21: Interim Operational Concept," April 1986, pp. 1-4 through 1-7; U.S. Army Training and Doctrine Command briefing, "AirLand Battle Future: Conceptual Underpinnings," 28 September 1990; and Headquarters, Department of the Army, Office of the Chief of Staff, *Louisiana Maneuvers—The First Year*, March 1994.

Although it draws on multiple intelligence and information sources, Army intelligence does not have primary responsibility for operating any of the main intelligence-collection disciplines: human intelligence, electronic intelligence, measure and signal intelligence, and so on. Its collection capabilities are focused on battlefield intelligence. As a result, Army intelligence must request support for its intelligence-collection plan from the organizations that operate the major collection efforts. Moreover, although Army intelligence devotes a large effort to production of futures intelligence, its investment in collection is focused in the present, in current intelligence. Thus, although the U.S. Army's Office of the Deputy Chief of Staff for Intelligence (ODCSINT) expends considerable effort on futures intelligence, it does not own the collection means to support this effort. It must rely on others for collection support.

A CLASH OF CULTURES

The cultures and operating modes of Army intelligence and long-range planners were—and remain—quite different. The strategic planners especially operated in a fast-paced mode in which their attention was often whipsawed across a wide range of issues. While one might assume that long-range planners would be insulated from the pace of current operations, that was rarely the case, even in the infrequent episodes when they were organized in separate, long-range planning divisions. Planners were routinely drafted to help handle other urgent tasks.[10] Moreover, the pace of their own work was influenced by the culture of ODCSOPS, where short-reaction-time requirements flourished. Thus, when they needed intelligence support, they needed it promptly (in days or weeks), before they moved on to other issues. In contrast, with many of its analysts insulated from the pressures of the Army Staff since the mid-1980s in field operating agencies located outside the Pentagon, intelligence pursued a busy but more measured agenda in which ODCSINT sought to develop formal intelligence-production requirements allowing Army intelligence to apply its assets efficiently in order to satisfy the greatest number of intelligence consumers with its for-

---

[10]Today in the Strategy, Plans, and Policy Directorate, for example, officers on the planning team have responsibilities for current, urgent actions in addition to their planning duties.

mally produced products. Before their release for official consumption, intelligence products had to be reviewed and approved to ensure their consistency with official intelligence community positions. Formal production (e.g., a published estimate) and the administrative process meant that important work was often available to the strategic planners only after their ability to apply it had passed.

Although ODCSINT has made progress with its "intelligence on demand" initiative (described later), Army intelligence remains product oriented, and many planners doubt their ability to keep up.[11]

THE EFFECTS OF DODFIP ON ARMY INTELLIGENCE

The DOD Futures Intelligence Program was intended to provide uniformity in forecasts, to dispose of multiple, duplicative projections, threats, and scenarios, and to make efficient use of intelligence resources.[12] A series of committees meets to determine what "products" the program will produce and to task the service and defense intelligence organizations. Figure 1 illustrates the overall DODFIP product line or "series architecture."

The category I products toward the top of the pyramid tend to be led by DIA. Service intelligence activities are consulted during the staffing and vetting—"coordination"—stage of development, but service intelligence organizations get to lead these efforts very infrequently. Indeed, the old strategic products Army intelligence once produced, the Long-Range Planning Estimate, Army Global Forecast, and similar publications, have all been supplanted by DIA products. The services fare somewhat better with category II products, although the J-2 and DIA remain very influential in this area. The services do much of the work in category III. This group of assessments

---

[11]See, for example, the planner workshop comments in Appendix A, "Current Problems with Intelligence."

[12]Interview with Ms. Christine McKeown, Defense Intelligence Agency, June 18, 1997. Also, DIA briefing, "Department of Defense Futures Intelligence Program," 11M9147-1/SWB, undated. See also *Department of Defense Intelligence Production Program: Special Production Programs*, DOD-0000-151E-96, chapter 2.

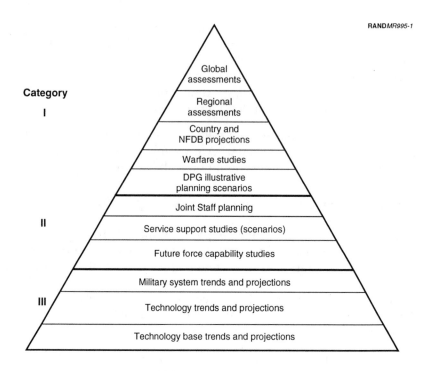

RANDMR995-1

Figure 1—The DODFIP Product Series Architecture

plays to each service's strength in, among other things, technical intelligence and foreign military equipment.[13]

Some DODFIP assets are well suited for general long-range planning. The National Futures Database, for example, maintains estimates of foreign capabilities for various points in time ranging from seven to twenty years into the future. Some of these data files are specifically designed to support acquisition planning, among other things.

Whatever its merits, DODFIP includes some drawbacks insofar as Army planning is concerned. In many respects, it is the perfect 1950s industrial-age organization. It emphasizes centralization and standardization, parsimony over innovation, and it places the service in-

---

[13]*DOD Futures Intelligence Program (DODFIP) Standard Operating Procedures for Generating Force Projections and Futures Studies,* DOD-0000-151E2-96, July 1996, p. 5.

telligence arms in order-taker mode. Although the services have some say in what they do, thanks to the committee structure that oversees the program, DODFIP distributes production requirements from all the services across all the military intelligence agencies to make efficient use of available assets. In a typical year, only about 8 percent of Army intelligence products are "initiative" products— intelligence produced because ODCSINT or one of its elements saw a need.[14] There is little flexibility left within Army intelligence to take on emerging questions, to address Army-unique perspectives, or to develop the case studies and scenarios necessary to support the Army Strategic Planning System. Indeed, if Army intelligence produced some of the scenarios strategic planners would probably want, ODCSINT could conceivably be guilty of violating the "lanes in the road" (the distribution of labor between the various intelligence agencies) that delineate DIA and service production responsibilities. The point is not to grant Army intelligence an unrestricted license to concoct its own scenarios that the service can use to compete for a larger budget share; but Army intelligence has legitimate responsibilities for protecting the Army from surprise and warning it of coming trouble, and it needs resources to work with planners on scenarios that fuel the planners' processes to do so. DODFIP behaves as if the Cold War intelligence production cycle were still in effect, requiring customers to submit requirements for intelligence and then tasking an agency to produce, coordinate, and release a product, rather than attempting "intelligence on demand," one of the objectives of Army intelligence XXI.[15] DODFIP is slow and formal, while Army planners are fast and informal.

Another shortcoming of the program is that Army intelligence is constantly struggling to get its products approved. Before the inception of DODFIP in 1993, some Army intelligence products did not require DIA approval and found acceptance among intelligence consumers. Since DODFIP, the percentage requiring approval has grown: virtually all ACQ work and most FD products now require DIA approval. In addition, although not attributable directly to

---

[14]Compiled from DIA's Consolidated Index of Intelligence Products (vol. 1, FY94, DOD-000-151G1-95).

[15]Army Intelligence XXI briefing, undated, seen at the National Ground Intelligence Center, June 19, 1997. .

DODFIP, more intelligence consumers seem to be insisting on fully vetted products. Despite the fact that the program includes a dispute resolution mechanism, DODFIP's emphasis on coordinated positions creates the danger that an "intelligence conventional wisdom" will develop, and that analytical approaches or conclusions that do not fit within it will be discounted.[16] Finally, DODFIP products seek to portray "most probable" estimates of the future, but Army strategic planning does not rely solely on probabilities; it monitors signposts derived from scenarios looking for empirical evidence of vulnerabilities undermining key assumptions.

CHAPTER OBSERVATIONS

This chapter has shown that intelligence support to long-range planning and long-range planning itself have had checkered and relatively short histories in the Army. Reinventing the Army has had a longer and relatively more successful run. At the same time, DODFIP poses a set of bureaucratic and methodological obstacles that complicate intelligence support to long-range planning. Can Army intelligence overcome these impediments? The answer depends in part on the specifics of long-range planning customer characteristics and needs, explored in the next chapter.

---

[16]Interviews at the National Ground Intelligence Center June 19, 1997.

Chapter Three

CHARACTERISTICS AND NEEDS OF LONG-RANGE PLANNING CUSTOMERS

STRATEGIC PLANNING

Support to strategic planning has an approach very different from that of supporting acquisition or force development. In the latter two types of planning, there is a legitimate need for point estimates because instructions and regulations from above the Department of the Army require them, and because they provide an agreed-upon set of performance characteristics that can be used as a standard across various acquisition and force development activities. Strategic planning, however, operates by identifying and monitoring the Army's key assumptions. The system uses scenarios and cases to try to stress these assumptions and cause them to fail. With knowledge of what would make an important assumption vulnerable, the system can monitor the landscape for signs of trouble.

Army assumptions may be vulnerable in ways that are not addressed in the formal, DOD scenarios like those found in the Defense Planning Guidance. Strategic planning must provide additional scenarios or case studies to cover the scenario domains left empty by official, fully coordinated scenarios, and to test critical assumptions exhaustively against vulnerability. Support to strategic planning means crafting the scenarios for assumption testing and subsequently monitoring the global security environment for indications that the Army's assumptions might be failing.

Strategic Planning Intelligence Needs

The strategic planners are concentrated in the Strategy, Plans, and Policy directorate of the Army DCSOPS. They also include the major command (MACOM) planners. The strategic planning problem set is qualitatively different from the one that confronts the acquisition and force development planners. The strategic planners want more than point estimates or consensus-based forecasts about the future. They seek evidence that suggests their important assumptions are becoming vulnerable.

Core Functions

At the highest level, Army strategic planners have two core functions: making sure the Army is adequately preparing for the future, and making sure the Army can do its part in the National Military Strategy. To accomplish these functions, strategic planners endeavor to

- help keep the Army substantially hedged against the most significant unfavorable outcomes; and

- prevent surprise.

Key Assumptions and Decisions

Our experience with Assumption-Based Planning indicates that an organization must discover its own load-bearing assumptions. Nevertheless, we believe some assumptions emerge from our workshops and interviews with officials as clearly important to the Army, and we offer them below. Key assumptions of strategic planners, as derived from workshop comments, focus on the following areas:

- the applicability of guidance contained in the National Security Strategy (NSS), National Military Strategy (NMS), and vision of U.S. warfighting;

- the state of the world, including the emergence of new major military powers;

- the nature of warfare; and

- resourcing (e.g., budget) levels.

Critical Signposts to Monitor

Among the signposts that DCSINT should monitor for Army strategic planners are:

- Changes in the nature of warfare;

- Changes in the current threat or opposing capabilities;

- Changes in the international environment;

- Changes in U.S. defense resources; and

- Changes in U.S. military capabilities.

ACQUISITION

Once threat-based (i.e., focused on Soviet military capabilities), the Army's acquisition practices have become more capabilities-based (i.e., focused on exploiting its own technological advantages).[1] Recent programs including the Crusader howitzer, Theater High-Altitude Area Defense (THAAD), the Comanche helicopter, the Javelin anti-tank missile, and product improvements to other major systems all reflect the influence of capabilities-based rather than threat-based development.[2] For example, Crusader was not envisioned as a response to developments in enemy artillery, but as the optimal exploitation of U.S. technological potential to produce a desired artillery capability. THAAD likewise sought to produce a desired air defense capability as a hedge against missile proliferation, even though the specific threat systems were held by only a handful of states. Comanche too, represents a desired capability to operate

---

[1]See HQDA, *Army Vision 2010*, especially the graphic on p. 10 that depicts "leveraging technology" as the foundation for "full spectrum dominance."

[2]This conclusion is based upon review of "Required Operational Capabilities" documents, STAR reports, and "Operational Requirements Documents." See, for example, HQDA ODCSINT, "Armored Gun System System Threat Assessment Report," June 1995, Headquarters, U.S. Army Aviation Center, "RAH-66 System Threat Assessment Report," November 1996, and HQDA ODCSOPS (DAMO-FDR), "Comanche RAH-66 Operational Requirements Document," January 1993. It is further supported by more general, official assertions that the Army does capabilities-based planning. See, for example, the *Army Science and Technology Master Plan,* Fiscal Year 1996, Volume I (Washington, D.C.: U.S. Government Printing Office, 1995), p. I-2, and *The Army Plan 1998–2013,* pp. 10–11.

forward of the FLOT (forward line of own troops), not as a response to enemy developments in armed reconnaissance capabilities. Javelin results as much from the age and limitations of the Dragon and TOW anti-tank weapons it replaces as from anything else. The Army sought a new, light, long-range anti-tank capability on its own merits, not in response to an emergent new armor threat. Requirements for new weapons and equipment reflect the changed demands of Army operational notions developed in the "Force XXI" and the "Army After Next" efforts rather than any specific threat. Except for certain legal requirements to produce system threat assessment reports (STAR reports) and similar documents for the acquisition process, Army intelligence's role has been diminished with the emphasis on capabilities-based planning.

There have been changes within DOD as well, including changing planner worries, changing acquisition practices, and downsizing of Army intelligence support. Although their predecessors tended to focus on specific "threat systems," planners today are at least as concerned about technological proliferation and hybridization and the trends in foreign technological development.[3] Planners want to assure the U.S. technological lead because it provides the freedom to base acquisition on U.S. capabilities and because it offers potentially high-payoff approaches.[4] Thus, despite downward trends in global defense outlays and arms transfers, planners need periodic reassurance that the U.S. lead in military technology is holding and that their acquisition strategy is not becoming vulnerable.

Acquisition (ACQ) Intelligence Needs

The Army acquisition community is an eclectic one, involving Army laboratories and arsenals, Program Executive Offices and Program Management activities for individual equipment and weapon systems initiatives, a host of commercial firms, and the directing and supervising offices of the Assistant Secretary of the Army for Re-

---

[3]Telephone interview with Colonel Steve Reeves, Chief Scientist's office within the office of the Deputy Assistant Secretary of the Army, Research and Technology, June 10, 1997.

[4]Planner comments from the project workshop at RAND, Washington, D.C., June 18, 1997.

search, Development, and Acquisition. Collectively this community seeks to create weapons and equipment that exploit the United States' strong suit in technology, while avoiding designs that would be vulnerable to foreign systems. Current policies can be safely characterized as "technology push"—exploiting U.S. technological advantages while watching for foreign technical developments and threats.

This community wants point estimates, "certified threats," formally published intelligence products, and consensus on future forecasts. Acquisition regulations—another Cold War legacy—demand these as a means to ensure that the equipment produced is appropriate for the enemies and conditions the Army is likely to encounter over a 20-year or longer time horizon: a daunting job for intelligence, indeed. Intelligence products, in this community, are often not only intended to inform the technical decisions, but are also used by the acquisition community to demonstrate compliance with laws and regulations.

Core Functions

At the highest level, ACQ's core functions within the Army are

- research, development, and acquisition (RDA) for new U.S. Army systems;

- RDA for modification of existing Army systems; and

- protection of individual Army systems from vulnerabilities, primarily to enemy systems (sometimes called weapon system life-cycle support).

Key Assumptions and Decisions

The key question that ACQ needs to have answered by DCSINT is "How could an adversary break or degrade current and planned U.S. Army capabilities?" DCSINT intelligence assessments provide the basis for ACQ planners' assumptions about the state of the world and emerging threats to Army capabilities. Three major sets of assumptions on factors that could break or degrade Army capabilities must be addressed by ODCSINT: assumptions about the importance of technology, assumptions about changes in adversary technologies, and assumptions about adversaries' ability to translate these technologies into militarily effective capability.

From our study of the capabilities-based approach, one unstated assumption of this community is that new technology—not asymmetric strategies, innovative doctrine and organization, or other nontechnological factors—will produce the most capable opposing forces or will be the principal source of capabilities that might "break" U.S. Army capabilities. Changes in adversary weapon systems can result from[5]

- indigenous technological breakthroughs;

- arms transfers;

- foreign technological exploitation; and

- co-production agreements.

Critical Signposts to Monitor for ACQ

These three sets of key assumptions appear to us to lead to a number of critical signposts that need to be monitored to ensure that Army capabilities are not negated.[6] These include data on the following:

- Estimates of current military and technological capabilities, and available resources

- Forecasts of changes in available resources

  — Among services, branches of services

  — To RDT&E, and to acquisition

  — To arms transfers, and to advisors

- Technological breakthroughs

  — Basic science or engineering breakthroughs

---

[5]Translation of technology into militarily effective *capabilities* requires operational concepts that exploit the technology by providing the enabling doctrinal, organizational, and training frameworks; these considerations appear to be given less emphasis by ACQ long-range planners than opposing technologies.

[6]All signposts derive from planner remarks during the project workshops.

— Product (weapon system) engineering and acquisition

- Technological diffusion (transfers of enabling technologies)

  — Arms transfers

  — Foreign technological exploitation

  — Co-production

- Incorporation of technologies into militarily effective capability

  — The emergence of asymmetric strategies

  — The emergence of new operational concepts

  — Innovative doctrine, organization and training

FORCE DEVELOPMENT

The force development process builds the force packages that provide the U.S. Army's deployable combat power and the combat developments activities that define the capabilities of the units within the force packages. Traditionally, intelligence support to FD meant providing approved threats and scenarios, developing foreign force structures and order-of-battle information, and describing foreign military operational practices. The task today has expanded. The FD community still wants to know the details of foreign militaries, but these details have expanded to scan more countries and to include more information about command, control, communications, computers, and intelligence systems (C4I). Force developers also need more intelligence about foreign communications networks and, in general, more about electronic infrastructure.[7]

But besides expansion of the traditional support role, other factors influence intelligence support to force development: the Army's capabilities-based approach, the focus on modern, conventional warfare as the principal test of force development initiatives, and planner worries.

---

[7]Interview with Major Scott Wilkerson, U.S. Army Concepts Analysis Agency, February 19, 1997.

The Capabilities-Based Approach

As was the case with acquisition, force development emphasizes capabilities over threat and other factors. Force XXI, the Army After Next, and their supporting activities—including the Battle Labs, advanced warfighting experiments, and similar initiatives—emphasize developing desirable capabilities to improve force effectiveness, efficiency, and the prospects of survival. As with acquisition, the capabilities-based approach appears to assume that technology push will have a major role in producing the best solutions that will be superior to those of foreign forces. Intelligence support must, therefore, monitor this assumption and help guard it from potential vulnerabilities. Monitoring this assumption means that Army intelligence must be on the lookout for foreign military developments that could cause it to fail.

The Focus on Modern, Conventional Warfare

The overwhelming majority of force development activities consider modern, high-tempo conventional combat operations to be the acid test. The advanced warfighting experiments at Fort Irwin pit the experimental force against a modern, mechanized enemy. The recent Army After Next Winter Wargame also pitted the United States against high-tech enemy forces.

With very little attention devoted to unusual modes of warfare, there is a danger that FD could be blindsided by adversaries who do not pursue a conventional way of war. For example, what if an opponent refuses to fight and simply hides its forces until the United States tires and withdraws? The United States would need the means to force a decisive engagement. Or suppose the opponent is an actor without obvious centers of gravity that U.S. forces are trained to locate and attack. The enemy may benefit from difficult terrain or have other advantages that allow them to fight in an unexpected way that keeps the Army from following its preferred mode of operations. Scanning the international landscape for potential foes who might practice unusual modes of warfare is thus a critical task in providing support to force development.

Planner Worries

A principal concern is developing intelligence against more of the credible regional actors, including nonstate, so-called transnational actors. Ideally, the FD community would wish to have files on these entities like the ones they maintain on states: equipment holdings, order of battle, and so on.

Asymmetric strategies—approaches to war that avoid U.S. strengths while concentrating on U.S. weaknesses—also enjoy a priority similar to that of tracking transnational actors. The planners in the project's workshops conceived of the asymmetrical strategy problem as distinct from unusual modes of warfare in that it emphasizes exploiting technology to gain advantage rather than rethinking warfare in some fundamental way. For example, an actor applying an asymmetrical strategy might have a thoroughly conventional military. His plan might be to embark on a terrorist campaign on the territory of U.S. allies, causing enough damage to fracture the coalition and deprive the United States of allied facilities while avoiding defeat on his own territory, maneuvering to stay out of the U.S. path until frustration caused a policy change and a U.S. withdrawal. As the example suggests, planners worry that rather ordinary forces might somehow be supplemented with limited amounts of equipment—terrorist bombs in this instance—that produce militarily significant results.

Force Development Intelligence Needs

Like their colleagues in acquisition, the force developers are engaged in exploiting technology to produce the most capable units possible and arraying them in force packages that yield the most versatile, lethal, and sustained land combat power possible.[8] Force developers watch foreign military developments for indications that they may make U.S. Army forces vulnerable.

Force developers also want point estimates, often provided as forecasted order-of-battle data on various foreign militaries. The community is seeking inputs for models and other analytical tools to help

---

[8]That is, within the constraints of the capabilities that have been developed by ACQ.

it understand the advantage of one type of unit relative to another, or one force package compared with another.

Core Functions

The core functions of Army force developers are

- comparative analyses of alternative force structures at all echelons; and

- identification of preferred force structures.

Key Assumptions and Decisions

FD needs the same question answered that ACQ does, namely "How could an adversary break, degrade, or limit U.S. Army current or planned capabilities?" and it has the same 20+ year time horizon as ACQ. Force developers do their planning on the basis of assumptions about likely modes of warfare and examine cases within and outside planning guidance parameters. The emergence of unanticipated modes of warfare or cases outside the guidance parameters can vitiate planning assumptions.

Unanticipated modes of warfare include

- unusual combat organizations;

- plans to achieve unique military or political objectives;

- innovative concepts of operation, use of technologies, or doctrine.

The emergence of cases outside planning guidance parameters can include

- cases in noncanonical regions;

- cases in unique environments (terrain, weather, etc.);

- restricted host nation support or base access;

- coalition constraints; and

- exercises or mobilizations that engender a new threat.

Critical Signposts to Monitor

Among force developers' long-range planning signposts that DCSINT should monitor are

- the effectiveness of foreign military organizations in the context of specific force-on-force scenarios;

- accuracy of portrayals of foreign capabilities;

- mobilization or force generation capabilities;

- changes in resources
  - budget allocations
  - RDA activities
  - arms transfers
  - advisors;

- major changes from current practices
  - unusual actors/methods of warfare
  - revolutionary organization or C2.

CHAPTER OBSERVATIONS

This chapter has summarized the characteristics and needs of the long-range planning customers. The strategic planners need broader types of information more applicable to their planning processes than what they currently receive. Two other intelligence consumers, ACQ and FD, continue to need Cold War–era point estimates for what seem to be legitimate, if somewhat dated, purposes. FD operates in an environment constrained by two potentially dangerous assumptions: that the capabilities-based approach is adequate and sufficient to field the best force for the future, and that a future peer competitor will remain the most difficult adversary. Even ACQ and FD, moreover, have sought information about trends in the world that could influence their activities. The question is, could Army intelligence deliver beyond its stock in trade of specific estimates? Does it have the tools to sort through the ever-growing mounds of data to produce the intelligence its long-range planning customers need? The next chapter considers the ability of technology to help.

Chapter Four

THE POTENTIAL OF INFORMATION TECHNOLOGY

Strategic planning in industry relies upon a wide range of processes and information systems that provide critical information. But there is a paradox with regard to investments in information technology:

- research has shown no relationship between corporate investments in information technology (IT) and profitability;[1]

- other studies suggest little relationship between national investments in IT and economic performance (growth in GDP);

- other research has shown that investments in information technology have led to little or no measurable increases in aggregate white-collar productivity in the United States.[2]

It seemed reasonable, therefore, to explore how successful corporate planners make use of IT. With this in mind, the study team sought to explore how industry thought about strategic planning and about how best to harness information technology in support of this planning, and what lessons might be useful to the DCSINT in improving support to Army long-range planners.

---

[1]According to Price Waterhouse Change Integration Team (1996, p. 162), few organizations have mastered the ability to develop systems that work, and IT project failure is an issue of great concern to project management:

Some surveys have indicated that as much as 75 percent of the money spent on new systems is expended on applications that either never make it into production or fail to meet the objectives that justified investment in the first place.

[2]McGee and Prusak (1993, p. 2), citing research by Stephen Roach and Gary Loveman.

COMMON THEMES ON INFORMATION SUPPORT TO STRATEGIC PLANNING

As specialists from Ernst and Young describe it, the fundamental problem for successful investments in IT remains embedding such decisions in an understanding of how technology can assist in providing "the right information, [at] the right time, and the right place":

> This is a definition that can be provided only by those executives charged with making consequential decisions for organizations. The answer, "all the information, right away, and everywhere," is untenable, no matter how often it is the implicit message in technology sales pitches or lazy information plans.[3]

Firms derive their information needs from the key performance measures that are used in strategic planning:

> Measurements—not data—are the foundation of management practice. Properly designed and used, measures can articulate strategy, drive change, shape behavior, focus action, and align management around activities that lead to success. Without sensible, balanced measurements, *most* of your organization's energy and actions are of no value to customers, to shareholders, or to employees. The worst measures (there are many) destroy value.[4]

These measures need to be aligned with business strategies and revisions to strategies, so that the firm can assess its current performance in terms of its goals and objectives while also assessing its ability to achieve the higher levels of performance envisioned in the next period. Put another way, the performance measures need both to reflect the performance objectives of the current period and to provide diagnostic information for the firm's ability to move to higher levels of performance.

These performance measures should also be balanced and small in number, focusing managers on a few key indicators that capture the essence of the organization's utility function. The identification of performance measures is best accomplished by filtering prospective

---

[3]McGee and Prusak (1993, p. xiv).

[4]Price Waterhouse Change Integration Team (1996), p. 236. Emphasis in original. For more on performance measurement, see Eccles (1991) and Kaplan and Norton (1992).

measures through the firm's objectives and strategies—if they don't appear to contribute to the bottom line in an important way, they are probably not the right measures.

Finally, as much attention is given to the process of developing performance measures as the measures themselves: it is essential that a consensus on performance measures be built among stakeholders, since a lack of ownership of such measures can cause them to be ignored or subverted. A Delphi or other group decisionmaking process is frequently used to this end, with participants asked to range candidate performance measures on the basis of

- relevance: the degree to which the measure is linked to the company's strategies and objectives;

- usefulness: how well the measure helps to identify the strengths and weaknesses of underlying business processes;

- understandability: how easily the measure can be understood; and

- availability of data: how easily the necessary data for the measure can be obtained.[5]

INDUSTRY CASE STUDIES

The three industries we chose to study were the computer manufacturing industry, the commercial aerospace industry, and the automobile industry (Table 2). These industries were selected because, like the Army, their firms typically have large budgets, make large investments in research and development (R&D), and operate long planning horizons.

Based upon the data in Table 3 we identified two companies— General Motors and Hewlett-Packard—that appeared to have particularly high absolute levels of R&D spending; with the merger of Boeing and McDonnell-Douglas, it seemed natural to choose Boeing for our commercial aerospace company.

---

[5]Price Waterhouse Change Integration Team (1996), p. 258.

Accordingly, we reviewed annual reports and other documents for these three companies, and we had a day of meetings with strategic planners in two of them.[6]

Table 2

Industries for Strategic Planning Analysis

Industry	Capitalization	R&D	Estimated Product Cycle	Estimated Life Cycle
Computer	High	High	9 months	3 years
Automobile	High	High	3–5 years	10 years
Aerospace	High	High	5–10 years	20 years

Table 3

Top R&D Spenders in 1996

Rank	Company	1996 R&D $ Millions	1995–1996 Change (%)
1	**General Motors**	**8,900**	**8.5**
2	Ford Motor	6,821	3.0
3	IBM	3,934	16.2
4	**Hewlett-Packard**	**2,718**	**18.1**
5	Motorola	2,394	9.0
6	Lucent Technologies	2,056	(23.7)
7	TRW	1,981	5.3
8	Johnson & Johnson	1,905	16.6
9	Intel	1,808	39.5
10	Pfizer	1,684	16.8

SOURCE: Data compiled by Andersen Consulting, 1997.

---

[6]Meetings with General Motors strategic planners were held on August 27, 1997, at their offices in Warren, Michigan; meetings with Hewlett-Packard were held on September 2, 1997, at their offices in Palo Alto, California. We were unable to arrange meetings with Boeing's planners but were sent information on the Boeing strategic planning process.

IMPLICATIONS FOR IDENTIFYING INFORMATION SUPPORT MODELS

Based on our literature review and interviews with corporate planners, Army long-range planning appears to have much in common with the activities that are intrinsic to strategic planning in industry.

A framework for applying what we learned from industry into the Army decision-based planning paradigm follows:

- *vision* and *objectives* guide the development of strategies;

- *strategies* systematically relate means to ends to realize the vision and achieve the objectives;

- *plans* capture the essential elements of strategies;

- *assumptions* are made that provide the premises for strategies and plans;

- *signposts* need to be designed to continuously test assumptions, to make sure that the assumptions (and plans) remain viable;

- *information needs* can be derived both from the signposts and through the mapping of key decisions that need to be made to their underlying criterion variables;

- *information support* should be designed to provide for information needs arising from the key assumptions and decisions; and

- *information delivery choices*—whether to use hard copy reports, briefings, information systems of various kinds, or other channels—should hinge on both the nature of the information and the nature of the audience, its needs, and its technological sophistication.[7]

COMPARISONS WITH THE ARMY

At systems level, the Army is making progress toward implementing its technical architecture. For example, the DISC4, Army Digitization

---

[7]For example, information systems can be designed either to give all users the same information or to allow users to establish profiles that filter through only the information of interest to them. More will be said of this later.

Office, and others have a mature plan for the system and technical architectures to support Army IT. Intelink, a secure portion of the Internet in which classified intelligence sites can communicate and interact, is but one product of this long-term effort, and currently it is available to about half of Army intelligence's consumers. About 30 percent of all Army intelligence is disseminated via electronic mail or by on-line data bases.[8] Intelligence on demand, an Army program just under way, seeks to make intelligence available to its consumers in electronic form as it is needed. Some progress toward better exploiting IT is being made. But problems persist.

As our case studies from industry noted, successful civilian planners make sure they understand customer needs before turning to IT. The Army has taken a similar approach, establishing the Intelligence Priorities Process Implementation Plan to create "a formal process for the Army to identify and voice its intelligence priorities in support of its Title X mission."[9] Army Priority Intelligence Needs Surveys (APINS) conducted in 1994 and 1995 provided a formal and high-priority means for intelligence consumers, including the long-range planners, to make their needs known. As the APINS responses made clear, however, ODCSINT did not have the resources to satisfy all customer requirements at once; the customers would have to prioritize their needs to assure that the most compelling were satisfied as soon as possible.[10]

Personnel reductions since 1991 have left ODCSINT and its field operating activities with 30–40 percent fewer analysts than before.[11] But the intelligence problem is more complicated than it was in 1991; when current and futures intelligence are considered together rather than concentrating on the Soviet Union as the principal threat, as the United States did through most of the Cold War, today's intelligence analysts find themselves facing growing lists of questions about the

---

[8]Army Priority Intelligence Needs Survey (APINS) (http://www/inscom.army.smil. mil/odcsint/update/current/initiate/apin/apin5.htm).

[9]Army's Intelligence Priorities Process Implementation Plan, 1994 (http://www. inscom.army.mil/odcsint/update/current/initiate/apin/apin7.htm). See also HQDA Action Memorandum Subject: Army Intelligence Priorities Process, February 7, 1994.

[10]Army Priority Intelligence Needs Survey (FY95) Completed (http://www.inscom. army.mil/odcsint/update/current/initiate/apin/apin5.htm).

[11]Estimate from ODCSINT.

world at large. Communications and connectivity technology offers at least the potential to help this smaller cadre of analysts satisfy the growing demands of various intelligence consumers.

Nevertheless, as the project team discovered during its workshops and from canvassing intelligence consumers, many customers, especially in ACQ and FD, remain wedded to paper products, which adds the printing process to intelligence production and creates an obstacle to the full exploitation of IT. Many, especially those involved in threat integration, are frustrated by a lack of timely responses from ODCSINT. Communications and connectivity technology has the potential to address some of these problems, especially if the planners can make the internal administrative adjustments necessary to accept electronic media "soft copies" instead of paper documents, largely a cultural adjustment. Just as the post exchange eventually came to accept credit cards in payment for goods, the various planning constituencies will eventually overcome their demands for paper products for some applications and accept soft copies. The important point is that ODCSINT must continue to prepare itself for the time—not far off—when intelligence on demand and proliferation of intelligence in cyberspace are the norm.

CHAPTER OBSERVATIONS

The Army's IT architectures are maturing and providing the means for easier, faster transmission of intelligence to the planning communities. The planners have some administrative hurdles to clear before they can fully exploit the speed and flexibility that will soon characterize intelligence on demand. Army intelligence must continue its IT efforts so that it is fully equipped to deliver the goods when the planners have overcome their administrative and cultural obstacles to cyberspace.

Chapter Five

POTENTIAL FOR INTELLIGENCE TO SATISFY
LONG-RANGE PLANNERS

This chapter considers the potential of Army intelligence to satisfy the intelligence requirements of three groups of long-range planners.[1] To arrive at an assessment of these prospects, the chapter considers the two principal groups of issues dealt with in this report, systemic issues and methodological issues.

SYSTEMIC ISSUES

Our research found three areas that suggest systemic problems: channels to customers, collection and analysis, and intelligence production direction.

Channels to Customers

These channels provide the connection between Army intelligence and its long-range planning customers. The channels to the strategic planners appear informal and fragile. They are based on the acquaintance of planners with intelligence officials and involve official but informal support: a quick paper here and there, some suggestions that planners read certain intelligence reports.[2] Although Army intelligence has provided limited amounts of formally staffed inputs to the strategic planners' documents like *The Army Plan*, we found

---

[1]This chapter draws on opinions expressed in interviews at the National Ground Intelligence Center, June 19, 1997, and on comments made during the April and June RAND workshops.

[2]Interview with Lieutenant Colonel Tim Daniels, ODCSOPS, February 19, 1997.

no intelligence production requirements written in direct support of strategic planners. Army intelligence does not appear to produce any intelligence product specifically for the strategic planners (although the planners make use of reports prepared for others). The channel to planners therefore appears rather limited and tenuous.

The channels to ACQ and FD both rely more heavily than the strategic planners do on formal intelligence products as described in Chapter Three. The informal, direct interaction with intelligence staff for ACQ has suffered with the reduction of foreign intelligence offices. The people at the remotest ends of the ACQ process, the contractors, must rely almost exclusively on products, since they have very little contact with analysts. Often the products available to the contractors and laboratories are little more than STAR reports. The channel to ACQ customers seems more clearly established than the channel to the strategic planners, but it appears to be no more robust.

FD may have the best, most robust connection, receiving an abundance of formal support while also maintaining direct contact with ODCSINT analysts. Well-established working relations with the War Plans Division of ODCSOPS and Concepts Analysis Agency provide both a paper trail of regular reports and force development activities in which Army intelligence participates, and also a face-to-face connection between ODCSINT staff and these customers. That said, it is interesting to note that the threat integration staff officers who were located within ODCSINT and who provided contact between ODCSINT and all the threat integration activities—and thus are part of the communications channel themselves—were among the most openly frustrated critics of Army intelligence during the workshops.

None of the channels between Army intelligence and its long-range planning customers are as sturdy as they might be.

Collection and Analysis

Two points emerged regarding collection and analysis. The first is that the amount of information collected across the nation's intelligence community is huge and that collection from national technical means is expensive. In the present era, where uncertainty predominates and intelligence must scan the world broadly for signs of trouble, the collection problem has expanded, producing more raw information from more sources that must be subjected to analysis to turn it into futures intelligence. Although, as noted in Chapter Two, much of its work is derivative of intelligence produced elsewhere, Army intelligence will nevertheless experience an increase in volume of information and intelligence that will prove taxing.

The second point is that Army intelligence does not have the number of analysts it once did. Downsizing has reduced the organization's capabilities at a time when collection is expanding to target a wider number of actors and regions around the world. Symptomatic of the lack of adequate analytical capacity is the organization's habit of detailing analysts away from their futures intelligence duties to staff the crisis of the day.[3]

Information technology is one possible source of help to make the analyst shortage less acute. Analysts need easy-to-use connections to the raw data and finished intelligence from other sources. They also need a rapid capability to search, organize, and correlate data from these materials. That said, long-range planners must also establish priorities so that the available resources can be applied against their most pressing issues.

Intelligence Production Direction

DODFIP controls most of Army intelligence's production, so if Army intelligence wants to fence some of its production resources to explore a potential threat, it must first convince DODFIP. To make its case, Army intelligence would have to be able to describe the potential problem in enough detail to convince DIA that it was credible, and that it posed a threat to all the services. But few potential problems—especially those in the future—present themselves initially in great detail. And some may be a major concern only to the Army. Finally, ODCSOPS' decision-based planning requires frequent assumptions testing, and often assumptions are not fleshed out with much detail. The present system for controlling intelligence production makes it very difficult for Army intelligence to secure the assets

---

[3]We are grateful to Bob O'Connell of the National Ground Intelligence Center for this observation.

necessary to investigate the typical potential problems and assumptions likely to interest Army strategic planners.

METHODOLOGICAL PROBLEMS

The study found two important methodological issues that bear directly on the prospects for Army intelligence to provide support to the long-range planning community: capabilities-based planning versus threat-based intelligence and the ability of the Army to anticipate.

Capabilities Versus Threat

As discussed in Chapter Three, planning remains capabilities-oriented, seeking to exploit U.S. technological advantage to develop the most capable ground forces possible. Army intelligence, with its broader view encompassing foreign military, political, economic, and societal-cultural factors, takes a very different approach to the future. The friction that results is valuable in at least two ways. First, Army intelligence's approach monitors the global security environment to assure the U.S. technological lead and thus the continued viability of capabilities-based planning. Second, it compels long-range planners to consider other (threat) factors that do not easily fit within their conceptions of planning. This is a healthy and valuable service for planners.

Anticipation

Anticipation lies at the core of long-range planning and futures intelligence. Planners need some sense of the circumstances and conditions likely to confront Army forces in the future so that they can frame the issues and recommend decisions about priorities among programs, budget guidance, and the like. Futures intelligence must be able to reach beyond current intelligence on military forces, weather, and terrain to offer planners some sense of the likely future in enough detail to be useful in framing issues, recommending priorities, and allocating budgets. Anticipation for Army intelligence involves the ability to manage surprise and warning, to spot innovation, and to manage organizational problems associated with working with planners.

Surprise and warning. Surprise and warning has historically fallen victim to a number of classical errors, described in Appendix B. Fundamentally, the surprise and warning problem for futures intelligence boils down to three questions: Are there indications of future trouble that Army intelligence can sense in the present, can their implications for the Army be accurately understood, and can Army intelligence convince the long-range planners and senior leadership, on the basis of the available evidence, that the indications mean what the intelligence officials claim they do? The problem includes some very demanding elements—knowing what indications to watch for, sensing them accurately, understanding their implications, and convincing the long-range planners and senior leadership that the interpretation offered is correct. The problem is further complicated because Army intelligence is not the sole source or authority on the global security environment and the state of international affairs. Planners have access to a wide variety of other sources, both official, like the U.S. Army War College Strategic Studies Institute and the Foreign Military Studies Office at Fort Leavenworth, and contract studies performed by professional service corporations and universities, to name just a few. The surprise and warning problem thus involves not only carefully sensing the indications that are available, interpreting them objectively and understanding what they mean for the Army, but competing against a multitude of other views for credibility with the planners.

Spotting innovation. Innovation has historically been an important indication of change and thus figures prominently in any process that seeks to prevent surprise and provide warning. At least five factors influence the prospects for spotting technological, and especially military, innovations. Two of these have to do principally with technological progress, while the others deal more with culture, government, and political forces that shape and temper innovation. The first of the technological progress factors has been called the "butterfly effect" or the "pinball effect."[4] At their essence, both no-

---

[4]James Burke, *The Pinball Effect* (Little, Brown, and Company: Boston, New York, Toronto, London, 1996).

tions point to the cumulative effects of individually insignificant events and suggest that historically, apparently inconsequential developments have often led to militarily significant advances. Put another way, both terms have to do with unexpected consequences: the pinball that ricochets about and sets other things in the game in motion; the butterfly that, despite its small size and weight, nevertheless sets off a chain of events by landing on something sensitive.

The second technological progress factor is time lag. In a study of U.S. military systems from conception to production, researchers found that "innovation events" generally culminated 20 years before the engineering design data that ultimately produced a piece of hardware.[5]

The other forces influencing the ability to perceive innovation include national assessments of security and political guidance (which, among other things, control whether certain states and actors can be considered potential adversaries and accounted for in long-range plans), and regional-cultural expertise that would enable observers to gauge a society's ability to innovate based upon its societal, cultural, economic, industrial, and educational capabilities and constraints.[6]

Managing organizational trouble. Army intelligence's suitability to support long-range planning is contingent in part on its ability to meet the demands for anticipation and forecasting of the planners and their planning system. Two classic errors—the quantitative fallacy, which focuses on quantifiable things solely because they are quantifiable, and the evidence gradient, which demands tougher standards of proof for undesirable outcomes—are more likely to lurk in official policies, such as approved methods for certain types of

---

[5]Raymond S. Isenson, "Project Hindsight: An Empirical Study of the Sources of Ideas Utilized in Operational Weapons Systems," in William H. Gruber and Donald G. Marquis (eds.), *Factors in the Transfer of Technology* (Cambridge, MA: MIT Press, 1969), p. 167.

[6]Paul A. Herbig, *The Innovation Matrix: Culture and Structure Prerequisites to Innovation* (Westport, CT: Quorum Books, 1994), M. Ishaq Nadiri and Seongjun Kim, "International Research and Development Spillovers, Trade and Productivity in Major OECD Countries," Working Paper 5801, National Bureau of Economic Research, Inc. (Cambridge, MA, 1996), and Barry R. Posen, *The Sources of Military Doctrine* (Ithaca, NY: Cornell University Press, 1984).

estimates, or in the form of official scenarios that exclude some issues as unrealistic (i.e., the evidence gradient to make them credible is too steep). It is difficult to determine the degree to which these factors manifest themselves in specific intelligence products, since the products are not uniform in their methodological approaches. The writing style often favors asserting intelligence judgments without supplying a comprehensive logic for arriving at them. The decision-based planning approach, by requiring key assumptions to be stressed in unusual ways, could help reduce Army intelligence's potential vulnerability to these factors.

The organizational problems compound the problem of credible anticipation and forecasting because the organization of defense intelligence reflects so many of them. Single, authoritative estimates, centralization, and intelligence shaped by policymakers—the key sources of trouble in organizational factors noted earlier—are the norm. If Army intelligence is to be successful in supporting long-range planning, it will need procedural and organizational room to move: freedom to depart from the organizational attributes that dominate most of defense intelligence.

Many of the tough issues surrounding technological innovation may be beyond Army intelligence's capability to deal with them. The problems of the pinball effect and time lag between innovation and actual design make direct monitoring very difficult indeed. In most instances they may defy observation, placing a premium on Army intelligence's established skills in forecasting and trend analysis as a substitute approach to trying to monitor larger numbers of specific episodes of innovation.[7] Another two problems, the content of national assessments and political guidance, are only marginally influenced by the Army. On the other hand, the ability to estimate how innovation will fare in a foreign state, and to appreciate how the state's societal, cultural, educational, economic, and industrial characteristics might support or impede innovation plays to the strong suit of Army intelligence's regional experts.

---

[7]Observation is further complicated by the fact that ODCSINT has more independent actors to watch. During the Cold War there were only a handful of troublemakers, with the Soviet Union at center stage. Today ODCSINT and its subordinate organizations must monitor over 60 countries.

CHAPTER OBSERVATIONS

At least three main points stand out. First, forecasting and trend analysis has an important role to play in compensating for the inability to monitor reliably modernization and innovation events. Second, Army intelligence's capabilities do not fully satisfy long-range planning customers in large part because the communications channels between them have atrophied. Finally, ODCSINT experts have the potential to make important contributions to the Army, but only if they can be successfully connected to their customers and only if they can somehow manage the large amounts of information they must process.

Chapter Six

CONCLUSIONS

This project sought to understand intelligence support to long-range planning, especially the principal sources of friction between Army intelligence and the long-range planning community. One thing that stands out clearly as a source of trouble between the two is that Army intelligence is threat based, while its customers, as the earlier chapters of this report make clear, are largely capabilities based. These very different orientations cause trouble because ODCSINT understands the threat to be a multidimensional problem involving political, economic, social, geographical, military, and technological factors, among others, while most planners currently focus narrowly on technology. Moreover, in capabilities-based thinking, the planners tend to concentrate their attention on their ability to exploit U.S. technology, remaining satisfied with periodic assurances from Army intelligence that the United States remains technologically dominant. The planners, although not ignorant of them, pay too little attention to other facets of the threat, for example, asymmetric strategies and unusual modes of warfare.

The planners seem to stake the Army's future on two important and potentially mistaken assumptions:

- Capabilities-based planning will produce superior forces.

- Preparing for a high-tempo, high-violence, force-on-force battle against a peer competitor is the most demanding challenge to confront the Army. If it can meet this test successfully, it can accomplish all other future missions.

The planners may be right, but the stakes are too high to allow betting or guessing. *ODCSINT should test these assumptions and provide independent answers.* The insights from such an intelligence effort should go far to inform acquisition, force development, strategic planning, and the ongoing reinvention efforts. Even if these assumptions prove to be entirely reasonable, their enduring validity is so vital to the Army that they bear careful monitoring by the intelligence staff for some time to come.

SOME SPECIFIC OBSERVATIONS

Based on our review of the evidence from the workshops, interviews, surveys, and other sources, we have a number of evaluative observations to make on intelligence support to Army long-range planning customers.

Strategic Planning

Support to strategic planners has been complicated by several factors that are well outside ODCSINT's span of control:

- The planners have been saddled with "trickle down" guidance from OSD and the Joint Staff that establishes a conventional wisdom about the future and limits the contributions and effectiveness of intelligence that is not congruent with the conventional wisdom. *Even though the planners have been saddled with this guidance, to keep the Army sufficiently hedged, ODCSINT needs to wrestle with less conventional conceptions of the future operating environment.*

- Army planners have not been engaged in strategic planning so much as programming, thus confusing themselves—and their intelligence providers in ODCSINT—about the true nature of their intelligence needs. Because of this, it is not entirely clear that the Army senior leadership is being informed by the types of long-range intelligence it needs to organize, train, and equip the Army for future operations. *ODCSINT should take a two-track approach, refining its understanding of the core strategic-level intelligence support needs of both the Army leadership and strategic planners, and develop capacity for delivering this information to both audiences.*

- Army strategic planners have historically been somewhat marginalized from higher-level Army decisionmaking, and their advice, therefore, has had at best only limited and indirect influence. Whatever the planners' fortunes, *ODCSINT should make it its mission to educate the senior leadership on the key intelligence and long-range planning issues—like the continued viability of capabilities-based planning and the use of a peer competitor as the most demanding test case for future field army designs.*

- DODFIP and the ODCSINT product orientation make it even more difficult for Army intelligence to support strategic planning. Because DODFIP exercises partial control over Army intelligence resources, ODCSINT has limited flexibility in assigning resources in response to Army planner needs. *ODCSINT might be able to make more effective use of available resources by developing less formal vehicles (not unlike the initial "flash" estimates on unemployment and inflation) that provide early, initial estimates with appropriate caveats about remaining uncertainties; information technologies might profitably be used both to disseminate this information and to make it seem less formal than a paper product.*

Acquisition

It appears that intelligence support to long-range planning in ACQ is working fairly well; while there will most likely be demands for additional intelligence from planners (outlined in Chapter Three), there is one systemic problem in evidence—poor intelligence-to-customer communications channels. STAR reports, frequently questioned during the project's workshops, serve a real purpose among laboratory and contractor personnel by providing data that serve at least as a starting point for considering the threat from foreign systems, even though the reports are often perceived solely as an administrative burden within ODCSINT. Complaints during the workshops about the timeliness and promptness of intelligence support suggest that ODCSINT lacks both a routine means to assess acquisition's changing intelligence needs and flexible organizations and systems for allocating resources to ensure responsiveness to these needs. Communications and connectivity technology and the intelligence on demand program should help address these concerns, but *only if ODCSINT works to ensure that the Army IT technical architecture specifically addresses the needs for connectivity between ODCSINT and its ACQ customers.*

Force Development

This group of planners could be great beneficiaries of Army intelligence. As noted in Chapter Three, throughout most of their reinvention efforts since World War II, intelligence and a wide array of other factors have played important roles. The narrowing focus of FD projects only became acute during the 1980s. *Army intelligence should aggressively seek to reestablish a close relationship with TRADOC, where the major FD reinvention initiatives typically take form. By demonstrating to the reinvention community the potential influences of the factors beyond technology that ODCSINT monitors— political, economic, cultural, and strategic influences, to name but a few—Army intelligence can prove its worth in these endeavors.*

Is Army Intelligence Broken?

Army intelligence is not broken, but it has serious deficiencies in its ability to support and to interact with its customers. At present, ODCSINT's channels of communications to its customers are badly underdeveloped. Army intelligence, like its counterpart intelligence agencies, is awash both in raw, unevaluated information and in intelligence, and it lacks adequate numbers of analysts and technical processing support to exploit fully all the information at its disposal. The logjam resulting from too much information and too few analysts is exacerbated by the insistence both within Army intelligence and among its customers on formal products that are fully staffed, vetted, and approved.

Despite its serious deficiencies in its ability to satisfy its long-range planning customers, Army intelligence performs a valuable service to the planners by assuring them of the continued viability of their preferred capabilities-based planning approach and by insisting that they periodically consider military, political, economic, and other factors that do not fit neatly within their preferred planning system.

A POSSIBLE STRATEGY: CHANGING THE WAY ODCSINT DOES BUSINESS

In changing the way it does business, ODCSINT should attempt to further refine its understanding of ACQ and FD futures intelligence needs, at the same time stepping into the breach to attend to strategic planning more directly, reestablishing a close relationship with the reinvention planners and making a more aggressive effort to exploit communications and connectivity technologies.

In changing its approach to strategic planning, Army intelligence should attempt to overcome the planners' fixation on programming issues by showing them the potential benefits of monitoring and studying the Army's most important assumptions. DCSINT should elicit from the Army senior leadership their strategic concerns and use them to fill out ODCSINT's strategic agenda for Army intelligence.

Over the long term, communications and connectivity technology also offers ODCSINT the potential to transform Army intelligence by linking intelligence experts with decisionmakers and their staffs. Army intelligence must have sustained, close interaction with the planners and senior leaders if it is to win the competition with other information sources for credibility and influence—the ingredients it needs to do its job of preventing surprise and providing warning to the Army. Better managing knowledge and intellectual capital— what experts know and where they can be reached—will be key. ODCSINT must understand that its experts, not its formal reports, are becoming its real products. Experts should contribute to solving the Army's problems based upon what they know rather than on where they sit in the organization chart. Information technology is not a panacea for the problems facing Army intelligence, but communications and connectivity technology can help provide the linkages so that senior leadership can reach the appropriate intelligence expertise no matter where the expert is assigned. ODCSINT can use communications and connectivity technology to accelerate the Army's movement toward becoming a knowledge-based organization, where knowledge and expertise on the intelligence staff become increasingly valued and better tools become available for tracking and using this knowledge.

Finally, as experts become more central to Army intelligence, ODCSINT should consider different professional development paths for them that allow them to deepen their expertise through the years. The foreign area officer program is an obvious pathway for regional experts, but functional experts also require a special career track if they are to be seriously regarded throughout the Army and across the defense establishment. Tours of duty with the Defense Advanced Research Projects Agency and similar assignments might provide useful experience for functional experts. ODCSINT should consider ways to manage its Military Intelligence officer personnel and civilian professionals to ensure that high-quality, credible expertise will always be available.

THE BOTTOM LINE

The ability to act on the foregoing suggestions is dependent on four things.

- ODCSINT must be able to fence some intelligence production assets for Army needs above those documented through the DODFIP if it is to work the types of Army-unique issues and scenarios discussed in this report. ODCSINT should negotiate with DODFIP to secure sufficient resources.

- Monitoring the Army's key assumptions is essential business in support of the planners. Whether long-range planners appreciate it or not, by monitoring the important assumptions, Army intelligence provides an umbrella under which the planners can pursue their narrower methodologies with some degree of safety and can have some warning if their preferred capabilities-based approach becomes vulnerable.

- Communications and connectivity technology is not a cure-all, but it plays an important role in connecting ODCSINT's smaller staff of experts with planners and other intelligence consumers. ODCSINT should make it a priority to work closely with DISC4 and the others involved to ensure the technical architecture will support the kind of connectivity Army intelligence wants and to broaden and deepen the channels to its long-range planning customers.

- It may be a truism that we live in a complicated world, but experts help make sense of it. ODCSINT must cultivate and grow the experts it needs in order to monitor assumptions and advise the leadership.

SOME CLOSING THOUGHTS

The job of Army intelligence has changed since the Cold War. No longer a case of "getting the Soviets right," it now revolves around performing the more demanding task of keeping the Army hedged against greater uncertainty so that it can always fulfill its responsibilities in the National Military Strategy. Doing so means taking a sharper and more catholic view of both the global security environment and the Army's intelligence needs, since not all the developments of concern to the Army will necessarily be visible in "trickle down" estimates and guidance drafted above the service. ODCSINT must be on the lookout for Army-unique threats that may lurk in domains not covered in illustrative planning scenarios or other high-level guidance. This is a tall order, but one to which ODCSINT must rise.

MAIN PLANNER CONCERNS FROM WORKSHOPS

Three one-day workshops were conducted. Each workshop contained 10–12 people drawn from some element of the long-range planning community, the Army Staff, and in at least one session, the Army Secretariat. Uniformed participants ranged in rank from major to colonel, while civil servants were generally in the GS-13 to GS-15 range.

Some workshop participants stood out as sharply critical of Army intelligence, characterizing it as "broken." As a result, we sought to gather additional views to see if this was a common perception, and we found that the pointedness of their charges was not echoed elsewhere. For example, none of the criticisms we heard during our e-mail canvassing of the acquisition community indicated that these sentiments were broadly shared. Similarly, the 1995 APINS found no support for the characterization of Army intelligence as broken; 100 percent of the respondents said they were getting at least some of the intelligence support they needed, and 51 percent perceived Army intelligence as improving in its accuracy, timeliness, quality, and responsiveness.[1]

---

[1]Army Priority Intelligence Needs Survey (APINS) (http://www/inscom.army.smil. mil/odcsint/update/current/initiate/apin/apin5.htm).

FIRST WORKSHOP

Key Questions

The focus of the first workshop was strategic planning. The planners identified the following as the most important questions to which they would need answers:

About resources:

- Has the incoming skill base of recruits significantly changed training requirements?
- Have budget levels damaged the Army's ability to expand?

About the nature of warfare:

- Are there viable asymmetric strategies that threaten U.S. interests?
- Can MOOTW be accomplished with forces designed for warfighting?
- Has information warfare become a viable military weapon?
- Is the United States still dominant in space?
- Is the offense still dominant?
- Does maneuver still predominate over fire power?

About the international environment:

- Beyond U.S. unilateral capabilities, does our alliance structure allow us to secure the full range of our interests?

About the threat:

- Will there be a threatening regional hegemon?
- Will the United States have a global peer competitor?

About capabilities:

- Are we still generally ahead in measures/countermeasures?

Current Problems with Intelligence

Planners quickly found consensus that Army intelligence is broken at a systemic level—organizational, process, products, and other factors all contribute to this. It needs to be rethought from the top down.

- The system is not organized, structured, or supplied with incentives to support long-range planning, and there are no easy solutions to making it more responsive to these needs. Intelligence supplied is not what is needed to operate, nor what is needed to plan.

- Title 10 (Army) needs differ from warfighters' needs in significant ways, but there is little recognition of this or its implications in the way the Army is organized for intelligence support to long-range planning.

- While it may be useful for commanders and others, much of the technology-heavy, expensive imagery and SIGINT is useless in intelligence support to long-range planning. Much better value for money could be had by taking some of that money and supporting more analysts/analysis, training and professional development, and support for tiger teams for crisis operations support.

- A narrow focus on canonical approved scenarios (the Defense Planning Guidance's Illustrative Planning Scenarios) is less useful than a more realistic menu of more likely ones.

- Organizational lines blur distinctions between support for crisis operations and support for planning—this means that the analysts who would be doing intelligence analysis for planning are always putting out the crisis du jour. The solution is organizational—isolate intelligence support to long-range planning from the daily fire drills.

- The focus of the Chief of Staff should be institutional and futures oriented, not getting up to speed on operations. Intelligence support should be helping to support the Chief's vision of the total Army for the future.

- Military Intelligence (MI) in particular needs to be rethought, reorganized. Are there too many MI units? Too many people in MI?

- There needs to be more interaction between planners and intelligence across the board, through serving on same working groups, cross-briefing of communities, and so on. Only this can sensitize intelligence staff to planners' needs, and planners to intelligence's capabilities.

- Planning needs to be more capabilities-based than it is now.

- In a world where Third World sideshows (peacekeeping, humanitarian, and similar operations) dominate, "intentions" are more important than "capabilities"—bad intentions and a low-tech sniper's rifle (or a radio-controlled mine) can ruin a commander's day, and compromise the political viability of the operation.

- HUMINT, SIGINT, and COMINT may be more important in a world of these Third World contingencies, because they can be revealing of "intentions."

- There are no mechanisms for institutional learning and training—when a new analyst comes in, he faces a clean slate. There need to be supporting systems that capture both substantive knowledge and indexed knowledge (e.g., directory of analysts/ offices that provide information, indexes and abstracts of relevant publications).

- In the absence of a threat, force structure is still organized around warfighting.

SECOND WORKSHOP

Key Questions

The second workshop was geared to acquisition and force development officials. Their key questions fell into five categories: geostrategic, technological, human-organizational, nature of warfare, and domestic.

Geostrategic:

- Has the United States been involved in a major theater war?

- Is the continuance of the United States as a global nation-state threatened?

- Have there been asymmetrical conflicts against the United States or its forces?
- Is resource scarcity a primary cause of conflict?
- Has there been a change in the global environment that dramatically affects warfighting?
- Has China emerged as a near-peer competitor?
- Have there been major global alignments or realignments?

Domestic:

- Has the United States been confronted with an insurgency at home?
- Is there domestic consensus on the U.S. military role in the world?

Nature of warfare:

- Has the United States been confronted with an insurgency at home?
- Has there been a change in the global environment that dramatically affects warfighting?

Technology:

- Has technology had a major impact on organizational structures and values?

OBSTACLES TO EFFECTIVE ANTICIPATION AND RESPONSE

This appendix draws heavily on the literature of warning and surprise and of military and technological innovation to assess Army intelligence's tools for supporting the Army Strategic Planning System. Warning, surprise, and innovation are good analogues to use to evaluate Army intelligence assets for their appropriateness in support of the proposed decision-based planning system, for several reasons. First, long-range planning is all about anticipating trouble and taking appropriate steps to respond to it. The warning and surprise literature studies the same problem while emphasizing the causes of failure in the past: the organizational, perceptual, and other factors that have allowed surprise and failed to produce credible warning. Second, ACQ and FD are especially concerned with technical and military innovation and the weapons and foreign military capabilities that such innovation may produce. The innovation literature should help establish the degree to which Army intelligence is likely to be able to improve its ability to spot signs of foreign innovation useful for ACQ and FD purposes.

CLASSICAL ERRORS

The literature suggests that certain factors have interfered with good warning and allowed surprise; still others make spotting innovation difficult.

Key Observations About Warning and Surprise

Synthesizing the most salient points from the warning and surprise studies suggests that at least seven major factors have injured the ability of military planners and their intelligence colleagues to avoid surprise and offer credible warning of enemy action.[1] These include:

Interpretive errors in which the psychological dynamics of individual, small group, and organizational behavior cause analysts to misinterpret the evidence. An example of this type of error from Army intelligence—but hardly unique to it—would be the Soviet Battlefield Development Plans,[2] which persisted in overstating Soviet capabilities. The Soviet experts responsible for much of the document were heavily indoctrinated in Soviet military prowess through the Army's foreign area officer program and as a result were often unable to interpret emerging evidence objectively.

Receptivity errors in which the "signal-to-noise ratio" surrounding the information makes its meaning difficult to understand, or the analyst's expectations are so disparate from the data that they cannot be accurately understood, or where the rewards and costs associated with making the assessment interfere (e.g., the data would lead to a conclusion so widely at variance with the conventional wisdom as to make the analyst appear foolish). Those familiar with Army planning in the mid-1980s for developing the Extended Planning Annex to the POM would agree that analysts persisted in exaggerated expectations of Army budget growth despite clear evidence and signals from the administration to the contrary, because doing so was consistent with local conventional wisdom and because it made otherwise impossible programs appear plausible.

---

[1] The factors summarized here are drawn from Yair Evron (ed.), *International Violence: Terrorism, Surprise and Control* (Hebrew University, Leonard David Institute, 1979), Klaus Knorr and Patrick Morgan (eds.), *Strategic Military Surprise: Incentives and Opportunities* (New Brunswick: Transaction Books, 1983), Ariel Levite, *Intelligence and Strategic Surprise* (New York: Columbia University Press, 1987), John Lewis Gaddis, "International Relations Theory and the End of the Cold War," *International Security* (Winter 1992/93), Vol. 17, No. 3, pp. 5–58, and Richard Betts, "Surprise Despite Warning," *Political Science Quarterly*, Vol. 95 (Winter 1980–81), No. 4, pp. 551–572.

[2] A product of the U.S. Army Intelligence and Threat Analysis Center during the 1980s.

The fallacy of the perfect analogy, in which an analyst concludes that because two cases share some similarities, they are in every way analogous.

The ethnomorphic fallacy, in which an analyst concludes that because U.S. culture perceives events a certain way, other cultures do as well. One example from Army intelligence involves a study of Soviet tanks from the mid-1980s that concluded certain tanks were inferior to their U.S. counterparts because they did not employ thermal-imaging sights, when in fact the Soviets were developing a different technology of equal sophistication. But the analysis assumed that because the U.S. solution called for thermal imaging, the Soviet solution would, too.

The quantitative fallacy, which elevates the value of facts in proportion to their susceptibility to quantification. Estimates that focused on the "bean count" of tanks and other primary weapons forecasted to be in the inventory of various foreign states typified the quantitative fallacy. Analysts counted what could be seen, without giving due attention to qualitative factors like the state of training and morale of the military forces, among other things. As a result, many Warsaw Pact forces and others were credited with greater capability than was ever the case.

The evidence gradient, where the more improbable or unwanted the event, the stiffer the tests of veracity and reliability for any evidence suggesting the event may be about to occur.

Organizational factors, where centralization, the use of single, authoritative estimates, and intelligence constrained by policy lead to inaccurate and faulty conventional wisdom. Trickle-down planning guidance typifies this error, in which the Army planners in the past found that successive layers of Defense Guidance, Joint Strategic Planning Estimates, and similar documents left them in a planning straightjacket.

SELECTED BIBLIOGRAPHY

Betts, Richard, "Surprise Despite Warning," *Political Science Quarterly* (Winter 1980–81), Vol. 95, No. 4, pp. 551–572.

Bowersox, Donald J., "The Strategic Benefits of Logistics Alliances," *Harvard Business Review*, July–August 1990, pp. 36–45.

Bleeke, Joel, and David Ernst, "The Way to Win in Cross-Border Alliances," *Harvard Business Review*, November–December 1991, pp. 127–135.

Burke, James, *The Pinball Effect*, Boston: Little, Brown, and Company, 1996.

Dewar, J. A., C. H. Builder, W. M. Hix, and M. H. Levin, *Assumption-Based Planning: A Planning Tool for Very Uncertain Times*, Santa Monica, CA: RAND, MR-114-A, 1993.

Eccles, Robert G., "The Performance Measurement Manifesto," *Harvard Business Review*, January–February 1991, pp. 131–137.

Evron, Yair (ed.), *International Violence: Terrorism, Surprise and Control*, Hebrew University: Leonard David Institute, 1979.

Fetterman, Roger, *The Interactive Corporation*, New York: Random House, 1997.

Fischer, Layna (ed.), *The Workflow Paradigm: The Impact of Information Technology on Business Process Reengineering*, Lighthouse Point, FL: Future Strategies, Inc., 1995.

Gaddis, John Lewis, "International Relations Theory and the End of the Cold War," *International Security* (Winter 1992/93), Vol. 17, No. 3, pp. 5–58.

Gascoyne, Richard J., and Koray Ozcubukcu, *Corporate Internet Planning Guide: Aligning Internet Strategy With Business Goals,* New York: Van Nostrand, 1997.

Herbig, Paul A., *The Innovation Matrix: Culture and Structure Prerequisites to Innovation,* Westport, CT: Quorum Books, 1994.

Isenson, Raymond S. "Project Hindsight: An Empirical Study of the Sources of Ideas Utilized in Operational Weapons Systems," in William H. Gruber and Donald G. Marquis (eds.), *Factors in the Transfer of Technology,* Cambridge, MA: MIT Press, 1969.

Kahaner, Larry, *Competitive Intelligence,* New York: Simon and Schuster, 1996.

Kaplan, Robert S., and David P. Norton, "The Balanced Scorecard—Measures That Drive Performance," *Harvard Business Review,* January–February 1992, pp. 71–79.

Knorr, Klaus, and Patrick Morgan (eds.), *Strategic Military Surprise: Incentives and Opportunities,* New Brunswick: Transaction Books, 1983.

Levite, Ariel, *Intelligence and Strategic Surprise,* New York: Columbia University Press, 1987.

McGee, James V., and Laurence Prusak, with Philip J. Pyburn, *Managing Information Strategically,* New York: John Wiley and Sons, Inc., 1993.

Pearce, John A., and Richard B. Robinson, *Formulation, Implementation, and Control of Competitive Strategy,* Chicago: Irwin, 1994.

Peters, John E., *The U.S. Military: Ready for the New World Order?* Westport, CT: Greenwood Press, 1993.

Pommert, Margaret, "Developing the Balanced Scorecard: Performance Measurement and Strategic Alignment," presentation to RAND Workshop Series on Organizational Innovation and Restructuring, July 10, 1997.

Prahalad, C. K., and G. Hamel, "The Core Competence of the Organization," *Harvard Business Review,* May–June 1990, pp. 118–125.

Price Waterhouse Change Integration Team, *The Paradox Principles: How High-Performance Companies Manage Chaos, Complexity and Contradiction to Achieve Superior Results,* Chicago: Irwin Professional Publishing, 1996.

Russo, William M., "Managing Change at Chrysler: A Voyage of Discovery," presentation to RAND Workshop Series on Organizational Innovation and Restructuring, July 31, 1997.

Sproull, L., and S. Kiesler, *Connections: New Ways of Working in the Networked Organization,* Cambridge, MA: MIT Press, 1993.

Taylor, Alex III, "The $11 Billion Turnaround at GM," *Fortune,* October 17, 1994, p. 54.